# LIFE'S LITTLE BERRY COOKBOOK:

## *101 Berry Recipes*

*by Joan Bestwick*

Life's Little Berry Cookbook
*101 Berry Recipes*

*by Joan Bestwick*

Copyright 2000
by Avery Color Studios, Inc.
ISBN# 1-892384-05-1
Library of Congress Card #00-30431
First Edition 2000
Reprinted 2006

Published by
Avery Color Studios, Inc.
Gwinn, Michigan 49841

Cover photo by Reg Lipp,
Falcon Photography, Marquette, Michigan

Proudly printed in U.S.A.

# Table of Contents

*I would like to dedicate this book:*

*to God and his blessings;*

*to my city slicker cousin Karen Huges and my Aunt Vivian Martell who are great advocates for my books;*

*and to my little brother John Manning who could never keep berries in any container.*

## *H*ello again,

Berries have always been in my life as a country girl. I grew up in Lachine, Michigan in an area where berries flourished and the berries we couldn't get in Lachine we drove up north to pick. So this was a perfect book for me to write. I've had a lot of experience using berries in my kitchen and I thought that I should share some of these delicious recipes. The taste of fresh picked berries bursting in your mouth is a hard earned pleasure, but the fruits of your labor are worth it, when all year long you can enjoy these little treats. Every family should experience picking. It can be a bonding experience and the results are great. So enjoy! May you live a simple life, full of love and God.

Joan

Blackberries are in season from July to the first part of October. They are black, long and slender, with hard seeds and very tasty. They are cultivated, but also grow in the wild. Blackberries have 84 calories per cup, and are full of potassium, Vitamin A and C and also have 46 mg of calcium per cup. To freeze, spread the berries in a single layer on the cookie sheet and freeze solid. Transfer frozen berries to plastic bags. They can also be canned following canner directions.

Blueberries are a relative of the cranberry and are native to the United States. Blueberries were probably the first snack food consumed in North America. Native Americans used to sun dry and smoke them to store them for the winter months. It is also related to the European bilberry and whortleberry. When looking for blueberries, they should be firm and dry. The skin should be smooth and deep purple-blue. Fresh blueberries can be stored from 10 and 14 days and they need to be washed before serving. Blueberries have 80 calories per cup. Blueberries provide about 5 grams of fiber per cup and are a source of Vitamin C. To freeze, spread the berries in a single layer on the cookie sheet and freeze solid. Transfer frozen berries to plastic bags. They can also be canned or dehydrated per directions.

Cranberries peak months for picking are September through January. Fresh cranberries can be put in a freezer bag and thrown in the freezer just the way they are. To use frozen cranberries, rinse in cold water and drain or partially thaw in the refrigerator. One pound of fresh berries measures 4 cups and makes about 1 quart of whole-berry or jellied sauce. Cranberries grow on a slender, trailing North American shrub. The shrub grows in damp ground and bears a tart red berry. The berries are the edible part of the plant. Cranberries have 46 calories per cup. Cranberries have Vitamin A and C.

Raspberries are a fruit that is closely related to the blackberry and loganberry. The fruit is of native European bush, usually red, occasionally white or black. It is renowned for its delicate flavor. Raspberries are ripe when they slip off the stem into your hand without resistance. During hot weather pick raspberries everyday or they will become overripe. Wait until the morning dew has disappeared before picking them. Dry fruit is less perishable than wet. When buying raspberries, avoid red-stained containers, they indicate overripe, wet fruit. Black raspberries or black caps are sweet and aromatic. They are full of Vitamin C and potassium.

Store berries in small shallow containers so that the weight of the top berries doesn't crush those underneath. To wash raspberries, place them in a colander and submerge twice in a sink full of cold water. Drain well. To freeze, coat a cookie sheet with a fine layer of non-stick spray, spread the berries in a single layer on the cookie sheet and freeze solid. Transfer frozen berries to plastic bags. For fresh tasting raspberries, thaw 1 pint of frozen berries in a syrupy solution of 2 cups water to 1/2 cup sugar. Drain and use immediately. Raspberries have 64 calories per cup. They are full of potassium, Vitamins A and B6.

Strawberries are a small fruit that is ground grown. They were eaten in Italy over 2000 years ago. Refrigerate fresh strawberries in shallow ventilated containers as soon as you pick them. Only wash the berries in cold water when you are ready to use them. Do not allow the berries to soak. Strawberries stored with the stems stay firm longer than those without stems. They will stay fresh for several days if you put fresh, unwashed berries in a container and top with a folded napkin. Cover the container and turn it upside-down and store in the refrigerator. Remove any bruised or damaged berries as soon as possible. To freeze whole berries, spread the berries in a single layer on the cookie sheet and freeze solid. Transfer frozen berries to plastic bags. You can freeze strawberries with or without sugar or in a sugar syrup. Strawberries have 48 calories per cup. They are full of Vitamin C and potassium.

# BLACKBERRIES

*When we are filled with pride,*
*we leave no room for wisdom.*

# Blackberry Sauce

*1 quart blackberries*
*1/2 cup hot water*
*1-1/4 cup sugar*
*dash of salt*
*dash of cinnamon*
*2 tablespoons cornstarch*

In a saucepan combine blackberries, water, sugar, salt and cinnamon. Bring to a boil. Reduce heat and simmer gently until it is slightly syrupy, about 5 to 8 minutes. Add 3 tablespoons water to the cornstarch, stirring to make a smooth paste. Blend this into the hot blackberry mixture. Cook stirring until mixture is slightly thickened and translucent, about 3-5 minutes. Cool. Serve cold over ice cream, biscuits and whip cream.

## Creamy Blackberries

*1 cup whipping cream*
*1/2 cup sugar*
*1/4 cup orange juice*
*3 cups fresh or frozen whole blackberries, thawed*
*4 slices of pound cake*

Whip the cream in a bowl until stiff; continue beating, gradually adding the sugar and juice until well mixed. Gently fold the berries into the cream mixture. Spoon onto each slice of pound cake.

## Fruit and Belgian Waffles

*1/2 pint fresh blackberries or 1 cup frozen blackberries, thawed*
*2 medium size nectarines, pitted and sliced thinly*
*6 tablespoons sugar*
*2 tablespoons fresh lemon juice*
*8 frozen waffles*
*1/2 cup heavy cream, whipped or 1 cup non-dairy whipped topping*

Place blackberries in a medium size bowl. Mash lightly with a fork. Add nectarines, sugar and lemon juice. Toss together until well combined. Cover with plastic wrap and set aside for 10 minutes or for up to 2 hours. Toast waffles until crisp. To serve place 2 waffles on each dessert plate, spoon on 1/4 of the fruit mixture, and top with whipped topping.

## Blackberry Cobbler

*1/2 cup butter or margarine*
*1 cup sugar*
*1 cup water*
*1-1/2 cup self-rising flour*
*1/2 cup butter or margarine*
*1/3 cup milk, room temperature*
*2 cups blackberries fresh or frozen, thawed*
*1/2 to 1 teaspoon cinnamon*
*2 tablespoons sugar*

Melt 1/2 cup butter in a 10-inch round or oval baking dish. Set aside. Heat sugar and water until sugar melts and set aside. Make cobbler dough by cutting butter into flour making fine crumbs. Add milk and stir with a fork until dough leaves the sides of the bowl. Turn out onto a floured board; knead 3 or 4 times. Roll out onto a 11 x 9 x 1/4" thick rectangle. Spread berries over the dough. Sprinkle with cinnamon and roll up like a jelly roll. Cut into 1-1/4 inch thick slices. Carefully place in the pan covered with melted butter. Pour sugar mixture carefully around slices. The crust will absorb the liquid. Bake for 1 hour at 350°. Fifteen minutes before removing from the oven, sprinkle 2 tablespoons of sugar over the crust.

# Blackberry Cake

**Cake:**
*2 cups sugar*
*1 cup butter*
*4 eggs*
*3 cups all-purpose flour*
*1 teaspoon cloves*
*1 teaspoon nutmeg*
*1 teaspoon cinnamon*
*1 teaspoon baking soda*
*1 teaspoon baking powder*
*1 cup buttermilk*
*1-1/2 cup fresh or frozen blackberries*

**Icing:**
*1 cup butter*
*1 pound box of confectioners sugar*
*1 teaspoon vanilla*
*3 tablespoons cold coffee*

Cake: In a bowl cream the sugar and butter together. Beat eggs and add to creamed mixture. Combine flour, spices, baking soda and powder. Stir into creamed mixture alternately with buttermilk. Carefully fold in the berries. Bake at 350° in three greased and floured 8 inch round pans for 30 minutes. Cool on a wire rack. Icing: beat all ingredients together until fluffy. Add more coffee if necessary. Spread frosting between cake layers, then sides and top.

# Blackberry-Pecan Pie

*Pastry for a single pie crust, pre-baked*
*4 cups fresh or frozen blackberries*
*1 cup sugar*
*1/3 cup quick cooking tapioca*
*1 teaspoon ground cinnamon*

**Pecan topping:**
*1/2 cup flour*
*1/4 cup sugar*
*1/4 cup cold butter*
*1/4 cup finely chopped pecans*

Preheat oven to 375°. In a medium saucepan, combine the blackberries, sugar, tapioca and cinnamon. Cook, stirring over medium heat until mixture starts to bubble. Reduce heat and simmer the berry filling, uncovered for 10 minutes, stirring frequently. Pour the berry filling into the pie crust. Top with pecan topping. Topping: In a medium bowl, combine flour, sugar and mix well. Cut the butter into flour mixture until crumbly. Add the pecans and stir gently. Bake for 30 minutes.

# BLUEBERRIES

*Gratitude is the sign of a noble soul.*

# Spiced Blueberry Jam

*1-1/2 quarts blueberries*
*2 tablespoons lemon juice*
*1/4 teaspoon cloves*
*1/4 teaspoon cinnamon*
*1/4 teaspoon allspice*
*1 package (1-3/4 oz.) powder fruit pectin*
*5 cups sugar*

Sterilize 8 (6-ounce) jelly jars; keep them in hot water until ready to fill. Wash and drain blueberries. Puree the blueberries in a blender or food processor. You need 1-quart of puree. In a large pan combine puree, lemon juice, cloves, cinnamon, allspice, and fruit pectin, stir to mix well. Over high heat, bring mixture to a full rolling boil; boil 1 minute, stirring constantly. Add sugar all at once. Again bring to a rolling boil; boil 1 minute, stirring constantly. Skim off the foam on the top of the jam. Ladle into hot, sterilized jelly jars and seal.

# Blueberry Peach Jam

*3 cups blueberries*
*2 cups peaches, pitted and peeled*
*1 package of fruit pectin*
*1/4 teaspoon cinnamon*
*7 cups sugar*

Crush blueberries and put in a medium saucepan. Cut the peaches into pieces and grind in a food mill or processor. Add to the blueberries. Mix fruit pectin and cinnamon into the fruit mixture and cook over high heat until mixture comes to a boil, stirring constantly. Add sugar all at once. Bring to a full boil and boil hard for 1 minute, stirring constantly. Remove from heat. Skim and stir to remove foam. Ladle into hot jars. Makes 8-8 ounce jars.

# Blueberry Marmalade

*1 medium orange*
*1 lemon*
*3/4 cup water*
*3 cups blueberries, crushed*
*5 cups sugar*
*3 ounces fruit pectin*

Remove the peel from the orange and lemon. Scrape excess white from the peel and cut into very fine shreds. Place in a large saucepan and add 3/4 cup water. Bring to a boil and simmer covered for 10 minutes, stirring occasionally. Remove the white membrane from the fruit and finely chop pulp, discarding the seeds. Add to the peel mixture along with the blueberries. Cover and simmer for 12 minutes. Add 5 cups of sugar and bring to a full rolling boil, boiling hard for 1 minute, stirring constantly. Remove from heat and immediately stir in fruit pectin. Skim off the foam, stir and skim for 7 minutes. Ladle into hot sterilized jars. Seal at once. Makes 6-1/2 pints.

## Blueberry Apple Conserve

*2 pints blueberries*
*5 cups coarsely chopped, tart apples*
*6 cups sugar*
*1 tablespoon grated lemon peel*
*1/3 cup lemon juice*
*1 cup slivered blanched almonds*

In a large saucepan, combine blueberries, apples, sugar, lemon peel and juice. Cook, stirring constantly over high heat until sugar is dissolved. Reduce heat; simmer, uncovered and stirring occasionally for 1-1/2 hours or until the mixture is very thick. Remove from heat and add the almonds. Sterilize 4-pint jars, leaving in hot water until ready to fill. Immediately ladle into hot jars, filling to within 1/2-inch from the top. Cap and seal at once.

# Blueberry Syrup

*2 cups blueberries*
*1/2 cup sugar*
*1/2 cup water*
*1 teaspoon lemon juice*

Place all ingredients in a heavy pan. Over medium-low heat bring to a simmer cooking for 10 to 20 minutes. Makes 2-1/2 cups.

# Blueberry Pudding

*1 quart blueberries*
*2 cups sugar*
*3 tablespoons butter*
*1/2 teaspoon salt*
*1/8 teaspoon cinnamon*
*3/4 cup water*
*1 loaf of bread*

Combine blueberries, sugar, butter, salt, cinnamon, and water in a saucepan. Bring to a boil and simmer for 10 minutes, stirring occasionally. Remove crust from bread and cut slices into 1-inch squares. Lay half of the bread squares on the bottom of a 9-inch square dish. Pour half of the hot blue-berry mixture over the bread. Repeat process with remaining bread squares and blueberry mixture. Chill for 2 hours. Serve plain or with non-dairy whipped topping. Serves 8.

# Blueberry Peach Trifle

*1-14 ounce can sweetened condensed milk*
*1-1/2 cups cold water*
*2 teaspoons grated lemon rind*
*1-3-1/2 ounce package instant pudding*
*2 cups whipping cream, whipped*
*4 cups pound cake, cut into 3/4-inch cubes*
*2-1/2 cups fresh peeled, chopped peaches*
*2 cups blueberries*

In a large bowl, combine condensed milk, water and lemon rind. Mix well. Add pudding and mix, beating until well blended. Chill for five minutes and fold in the whipped cream. Spoon 2 cups pudding mixture into a 4-quart glass serving bowl and top with 1/2 of the cake cubes, all of the peaches, 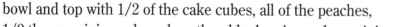 1/2 the remaining cake cubes, then blueberries and remaining pudding mixtures, spread to within 1-inch of the edge of the bowl. Chill for at least 4 hours. Makes 20 servings.

## Chilled Blueberry Soup

*2 pints blueberries*
*3/4 to 1 cup maple syrup*
*2 tablespoons lemon juice*
*1 tablespoon lime juice*
*1 cup buttermilk or half and half or light cream*

In a larger saucepan combine berries, maple syrup, lemon and lime juice. Simmer mixture uncovered for 30 minutes. Puree the soup in a blender until smooth and chill. Combine chilled soup to cream or buttermilk or half and half. Chill for about 4 hours.

## Blueberry Salad

*1 cup sour cream*
*2 tablespoons corn syrup*
*1 teaspoon lemon juice*
*1/4 teaspoon nutmeg*
*1/4 teaspoon cinnamon*
*2 cups blueberries*
*2 cups diced cantaloupe*
*2 cups seedless green grapes, halved*

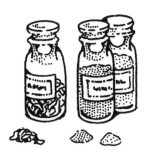

In a small bowl, combine sour cream, corn syrup, lemon juice, nutmeg and cinnamon. Mix well. In a medium bowl, mix the fruit together. Pour the sour cream mixture over the fruit and mix well. Chill.

## Blueberry Orange Salad

*2 cups orange juice*
*1-6 ounce package orange flavored gelatin*
*1/4 cup sugar*
*1 teaspoon grated lemon rind*
*2 cups buttermilk*
*2 cups fresh blueberries*
*lettuce leaves*

Bring orange juice to a boil in a saucepan over medium heat. Remove from heat; add gelatin, sugar and lemon rind. Stir in buttermilk and mix well. Fold in blueberries and pour into an oiled, 6 cup gelatin mold. Refrigerate until set. To serve, unmold onto lettuce leaves onto a serving platter.

## Blueberry Flapjacks

3 eggs
1 cup sifted flour
3 teaspoons baking powder
1/2 teaspoon salt
2 teaspoons granulated sugar
1 teaspoon light brown sugar
1/2 cup buttermilk
2 tablespoons butter melted
1-1/2 cup blueberries

In a large bowl beat eggs until light and fluffy with a mixer on high speed, about 2 minutes. Add flour, baking powder, salt and granulated sugar. Add the brown sugar and beat until smooth. Stir in the buttermilk and butter just until combined. Do not over beat. Fold in the blueberries. Slowly heat a well oiled griddle or skillet. To test the temperature of the pan drop some water on the pan and if it is hot enough the water should roll off in droplets. Use 1/4 cup of batter for each flapjack. Cook until bubbles form on the surface and the edges become dry. Turn and cook 2 minutes longer or until browned and serve.

# Blueberry Lemon Bread

**Bread:**
1/3 cup butter or margarine, melted
1 cup sugar
3 tablespoons lemon juice
2 eggs
2-1/2 cups all-purpose flour
1 teaspoon baking powder
1/2 teaspoon salt
1/2 cup milk
2 tablespoons grated lemon peel
1/2 cup chopped nuts
1 cup fresh or frozen blueberries

**Glaze:**
2 tablespoons lemon juice
1/4 cup sugar

In a mixing bowl, beat butter, sugar, juice and eggs. Combine the flour, baking powder and salt; stir into egg mixture alternately with milk. Fold in the peel, nuts and blueberries. Pour into a greased 8 x 4 x 2 inch loaf pan. Bake at 350° for 60-70 minutes or until bread tests done. Cool in pan for 10 minutes. While bread is baking combine the glaze ingredients. Remove the bread from pan and drizzle with glaze. Cool before serving.

## Blueberry Muffins

*1 cup fresh or frozen blueberries, thawed*
*1 tablespoon plus 1-3/4 cup flour, divided*
*1/2 cup sugar*
*1 teaspoon baking powder*
*1/2 teaspoon baking soda*
*1/2 teaspoon ground nutmeg*
*3/4 teaspoon salt*
*1 egg*
*1 cup sour cream*
*1/3 cup milk*

Preheat oven to 400°. Grease twelve muffin cups. In a small bowl toss blueberries with 1 tablespoon of flour and set aside. In a large bowl, combine the 1-3/4 cup flour, sugar, baking powder, baking soda, nutmeg and salt, set aside. In a medium bowl beat egg and stir in the sour cream and milk. Stir this mixture into the flour mixture until just combined. Batter will be lumpy. Stir in the reserved blueberries just until mixed. Fill muffin cups 2/3 full with batter. Bake until gold brown, about 20 minutes.

## Blueberry Oatmeal Muffins

1/3 cup flour
1 cup oatmeal, uncooked
1/4 cup brown sugar
1 tablespoon baking powder
1/2 teaspoon salt
1/2 teaspoon cinnamon
1 cup milk
2 egg whites
3 tablespoons canola oil
1 teaspoon vanilla
1 cup blueberries
2 tablespoons sugar

Coat 12 muffin tins with cooking spray. In a large bowl, combine flour, oatmeal, brown sugar, baking powder, salt and cinnamon. Make a well in the center of this mixture. Add the milk, egg whites, oil and vanilla. Stir until just moistened and fold in the blueberries. Fill muffin tins 2/3 full and sprinkle the tops with the 2 tablespoons sugar. Bake at 425° for 20 minutes or until golden brown.

# Blueberry Orange Muffins

**Muffins:**
*1 cup quick cooking rolled oats*
*1 cup orange juice*
*1 teaspoon grated orange zest*
*1 cup vegetable oil*
*3 eggs, beaten*
*3 cups all-purpose flour*
*1 cup sugar*
*4 teaspoons baking powder*
*1 teaspoon salt*
*1/2 teaspoon baking soda*
*3-4 cups fresh blueberries*

**Topping:**
*1/2 cup finely chopped nuts*
*3 tablespoons sugar*
*1/2 teaspoon cinnamon*

Preheat oven to 400°. Mix oats, orange juice and zest. Blend in the oil and eggs and set aside. Stir together the flour, sugar, baking powder, salt and baking soda. Add to the oat mixture and mix lightly. Fold in the blueberries. Spoon batter into paper lined muffin tins, filling 2/3 full. Combine the topping ingredients and sprinkle over the batter. Bake for 15-20 minutes or until lightly browned. Makes 24 large muffins.

# Blueberry Bran Muffins

*1-1/2 cup bran cereal*
*1 cup buttermilk*
*1 egg, beaten*
*1/4 cup melted butter*
*1 cup flour*
*1/3 cup brown sugar*
*2 teaspoons baking powder*
*1/2 teaspoon baking soda*
*1/2 teaspoon salt*
*1 cup blueberries*

Preheat oven to 400°. In a medium bowl, combine the bran cereal and buttermilk, let stand until the liquid is absorbed. Then stir in the egg and butter and set aside. In a different bowl, stir together the flour, brown sugar, baking powder, baking soda and salt. Add the bran mixture and stir together until mixture is just moistened. Fold in the blueberries. Fill 12 greased muffin cups 2/3 full. Bake for 20-25 minutes. Makes 12 muffins.

# Blueberry Crisp

*3 pints blueberries*
*1/2 cup sugar*
*juice and finely grated zest from 1 lemon*
*3 tablespoons flour*

**Topping:**
*1 cup flour*
*1 cup packed light brown sugar*
*2/3 cup rolled oats*
*1/2 teaspoon cinnamon*
*1/4 teaspoon salt*
*10 tablespoons cold, unsalted butter, cut into small pieces*

Preheat the oven to 400°. Butter a 9 x 13 inch glass pan. Mix the berries, sugar, lemon juice, zest and flour in a large bowl. To make the topping, combine the flour, brown sugar, oats, cinnamon and salt in the bowl of a food processor. Add the butter and pulse the processor repeatedly in 2-3 second bursts until the mixture is clumpy. Transfer the berries to the baking dish and spread the topping evenly over the fruit. Bake for 30 minutes, until hot and bubbly.

# Blueberry Coffee Cake

**Cake:**
*1/2 cup shortening*
*1 cup sugar*
*2 eggs*
*1 teaspoon vanilla*
*2-1/4 cups flour*
*1 tablespoon baking powder*
*2/3 cup milk*
*2 cups fresh or frozen blueberries*

**Topping:**
*2/3 cup sugar*
*2/3 cup flour*
*1/2 cup margarine*
*1/4 teaspoon almond extract*

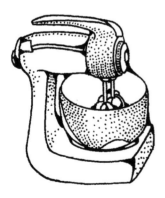

Beat the shortening and sugar until light and fluffy. Add the eggs and vanilla, beating well. Stir together the flour and baking powder and add to beaten mixture, alternating with the milk. Fold in the blueberries and spread in 2 greased (9-inch) round baking dishes. Combine topping ingredients and cut until crumbly. Sprinkle over the blueberries. Bake at 350° for about 35 minutes. Cool slightly and cut into wedges.

# Blueberry Oatmeal Cake

**Cake:**

*1-1/3 cup flour*
*3/4 cup quick cooking rolled oats*
*1/3 cup sugar*
*2 teaspoons baking powder*
*1/4 teaspoon salt*
*1 egg*
*3/4 cup milk*
*1/4 cup cooking oil*
*1 cup blueberries*

**Icing:**

*1/2 cup powdered sugar*
*2 to 4 teaspoons milk*

Grease an 8x8x1-1/2 inch round cake pan. In a bowl stir flour, oats, sugar, baking powder and salt and set aside. In a medium mixing bowl, stir together egg, milk and cooking oil. Add all at once to the dry mixture, stirring until just moistened, batter should be lumpy. Fold in the blueberries. Spoon batter into baking pan and bake at 400° for 20 to 25 minutes or until cake tests done. Cool 5 to 10 minutes. Drizzle cake if desired with powered sugar icing. Icing: in a small bowl, mix powered sugar and milk. Mix to drizzle consistency.

# Blueberry Pudding Cake

**Pudding Cake:**

*2 cups flour*
*1-1/2 cups sugar*
*2 teaspoons baking powder*
*3/4 teaspoon nutmeg*
*1/2 teaspoon salt*
*1/2 teaspoon grated lemon peel*
*3/4 cup butter softened*
*2 eggs*
*3/4 cup milk*
*2 cups blueberries*

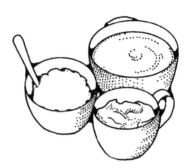

In a large bowl, make pudding cake by combining flour, sugar, baking powder, nutmeg, salt and peel. Then cut in the butter with a pastry blender until it resembles small peas. Add eggs and milk and mix on low speed for 3 minutes. Pour into greased and floured 9x9x2 inch pan. Top with the blueberries. Bake at 350° for 1 hour and 10 minutes or until tester inserted in the center comes out clean.

# Blueberry Pudding Cake, Cont.

**Lemon Sauce:**
*1/2 cup sugar*
*1 tablespoon cornstarch*
*1/4 teaspoon salt*
*1/4 cup cold water*
*1 egg yolk*
*3 tablespoons lemon juice*
*1 teaspoon grated lemon peel*
*2 tablespoons butter*

To make the sauce combine sugar, cornstarch and salt in a saucepan. Stir in cold water, cook and stir over medium heat for 10-12 minutes or until clear and quite thick. Blend the egg yolk with the lemon juice and gradually stir into the sauce. Stir in peel and butter. Serve the warm pudding cake with the sauce. Makes 9 servings.

## Blueberry Peach Pie

*2 tablespoons lemon juice*
*3 cups peaches, peeled, pitted and sliced*
*1 cup blueberries*
*1 cup sugar*
*2 tablespoons quick cooking tapioca*
*1/2 teaspoon salt*
*Pastry for a two crust pie*
*2 tablespoons butter*

Preheat oven to 425°. In a large bowl stir lemon juice and fruit. Add the sugar, tapioca and salt. Mix until combined then let stand for 15 minutes. In a 9-inch pie pan, place 1/2 of the pie pastry. Add the fruit mixture mounding in the center. Dot with butter, top with pie crust, seal and make several slits in the top. Bake for 40-45 minutes or until the crust is golden brown. Cool.

# Blueberry Sour Cream Pie

**Pie:**

*1 unbaked 9-inch deep dish pie crust*
*3 cups blueberries*
*1 cup sugar*
*1/3 cup all-purpose flour*
*1/8 teaspoon salt*
*2 eggs*
*1/2 cup sour cream*

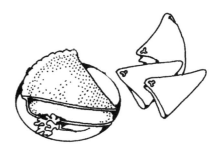

**Topping:**

*1/4 cup softened butter*
*1/2 cup sugar*
*1/2 cup flour*

Pour fruit into the pie shell. In a bowl mix remaining ingredients and pour on top of the blueberries. Topping: mix and sprinkle the crumble topping over the berry mixture. Bake at 350° for 1 hour.

## No Bake Blueberry Pie

*1 9-inch pie shell, baked*
*3/4 cup sugar*
*2-1/2 tablespoons cornstarch*
*1/2 teaspoon salt*
*2/3 cup water*
*2 pints blueberries*
*2 tablespoons margarine*
*1-1/2 teaspoon lemon juice*
*non-dairy whipped topping*

In a medium pan combine sugar, cornstarch and salt. Add water and 1 pint of the blueberries. Bring to a boil, stirring until thick. Stir in the margarine and lemon juice and cool. After it has cooled to room temperature, fold in the other pint of berries. Spread a layer of non-dairy whipped topping over the bottom of the shell. Add the berry mixture to the pie and serve with additional non-dairy whipped topping.

# CRANBERRIES

*Sacrifice is the true measure of our giving.*

## Cranberry Chutney

*1 large can (1 lb. plus) pineapple chunks*
*2 cups sugar*
*1 pound fresh cranberries*
*1 cup white raisins*
*1/2 teaspoon cinnamon*
*1/2 teaspoon ginger*
*1/4 teaspoon allspice*
*1/4 teaspoon salt*
*1 cup walnuts, broken*

Drain the pineapple, reserving the juice; set the fruit aside. In a medium saucepan, combine the pineapple juice, sugar, cranberries, raisins, spices and salt. Cook until the mixture boils. Lower the heat and simmer for 25 minutes. Add the pineapple and nuts and mix well. Remove the pan from the heat, cool and store in the refrigerator.

## Cranberry Relish

*1 pound cranberries, fresh or frozen*
*2 cups sugar*
*1 orange, unpeeled*
*2 apples, cored and unpeeled*

With a meat grinder, grind the cranberries. In a bowl mix the cranberries with the sugar and let this mixture sit. Meanwhile, finely chop the orange and apples. Add this to the cranberry mixture and mix well. Chill overnight. Makes 4 cups.

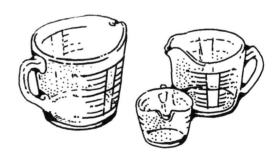

## Cranberry Vinegar
**(Perfect for a salad dressing, poultry and meat marinades.)**

*1 cup cranberries, fresh or frozen*
*2-6-inch bamboo skewers*
*3-1/2 cups white vinegar or white wine vinegar*
*1/2 cup mild flavored honey*

Wash and pick over the cranberries. Thread a few cranberries onto two bamboo skewers. In a medium saucepan, heat cranberries and vinegar. Bring to a full rolling boil. Cranberries should begin to pop. Remove from the heat and set the mixture aside. Into this mixture, stir in the honey to blend. Pour hot vinegar into two sterilized pint jars or bottles. Add a skewer of cranberries to each jar and seal.

# Frozen Cranberry Salad

*1-16 ounce can cranberry sauce*
*1 cup canned apple pie filling*
*1 cup sour cream*
*1 tablespoon lemon juice*
*3 tablespoons powdered sugar*
*lemon slices and fresh cranberries for garnish, optional*

In a mixing bowl, gently mix together cranberry sauce and apple pie filling. Combine well, do not mash the apples. In another bowl, mix the sour cream with the lemon juice and sugar. Stir into the cranberry mixture and spread into a foil lined 8-1/2 x 4-1/2 inch loaf pan. Cover and freeze salad for at least 4 hours or until ready to serve. Remove from the pan onto a serving platter. Let stand at room temperature for 5 to 10 minutes. Garnish with lemon and fresh cranberries if desired.

# Cranberry Coleslaw

*1 cup fresh cranberries, chopped*
*1/4 cup sugar*
*3 cups finely shredded cabbage*
*1/2 cup orange juice*
*2 tablespoons chopped celery*
*2 tablespoons chopped green pepper*
*1 cup green grapes, seeded and halved*

**Dressing:**
*2 tablespoons mayonnaise or salad dressing*
*2 tablespoons plain yogurt*

Combine the cranberries and sugar in a small bowl and set aside. Combine the cabbage with the orange juice in a large bowl. Add the cranberries and sugar, celery, green pepper and grapes, toss. In a small bowl, combine dressing ingredients. Again toss the dressing mixture with salad. Chill before serving.

## Cranberry Filled Acorn Squash

*1 medium acorn squash*
*1-1/2 cups fresh cranberries*
*6 tablespoons apple juice*
*1/4 teaspoon ground nutmeg*
*1/8 teaspoon ground cloves*
*1 teaspoon cornstarch*
*1 tablespoon walnuts, chopped*

Cut the squash in half and remove the seeds. Take plastic wrap and cover each half. Microwave on high for 6-9 minutes or until soft. Remove the squash from the microwave. In a small microwave casserole dish, combine the cranberries, 1/4 cup apple juice, sugar and spices. Cover and microwave on high for 4 to 6 minutes, until the cranberries pop. Combine the remaining apple juice with the cornstarch in a bowl and mix well. Gradually stir this into the cranberry mixture, mix well. Return to microwave and cook on high for 45 seconds to 1-1/2 minutes, until thickened. Spoon cranberry mixture into squash shells, sprinkle with the chopped walnuts and microwave on high for 1 to 2 minutes. Makes 4 servings.

# Cranberry Stuffing

*1 pound bulk Italian sausage*
*1/2 cup celery, chopped*
*1/4 cup onion, chopped*
*1 egg*
*1-8 ounce package herb seasoned stuffing mix*
*1 cup chicken broth*
*1/2 cup fresh or frozen cranberries, chopped*
*1/3 cup butter*

Preheat the oven to 325°. In a medium non-stick skillet over high heat, add sausage, celery and onion. Cook until meat is browned and vegetables are softened. Remove from the heat and drain. In a bowl beat the egg then stir in sausage mixture, stuffing mix, broth and cranberries. Place mixture in a greased bread pan and drizzle with butter. Bake until browned, 30 to 40 minutes.

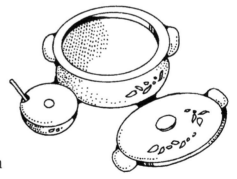

## Cran-Twist Turkey Sandwiches

*2 cups cooked turkey, cut into cubes*
*1/2 cup mayonnaise*
*1/2 cup fresh or frozen cranberries, finely chopped*
*1 orange, peeled and chopped*
*1 teaspoon sugar*
*1 teaspoon prepared mustard*
*1/2 teaspoon salt*
*1/4 cup pecans, chopped*
*lettuce leaves*
*6 rolls or croissants, split*

In a medium bowl, combine the turkey, mayonnaise, cranberries, orange, sugar, mustard and salt. Just before serving, stir in the pecans. Place lettuce and 1/2 cup turkey mixture on each roll. Serves 6.

## Cranberry Chocolate Ice Cream

*2 cups fresh or frozen cranberries*
*1/2 cup orange juice*
*1/2 teaspoon almond extract*
*1 quart vanilla ice cream, softened*
*3/4 cup semisweet mini chocolate chips*

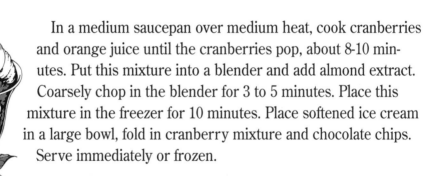

In a medium saucepan over medium heat, cook cranberries and orange juice until the cranberries pop, about 8-10 minutes. Put this mixture into a blender and add almond extract. Coarsely chop in the blender for 3 to 5 minutes. Place this mixture in the freezer for 10 minutes. Place softened ice cream in a large bowl, fold in cranberry mixture and chocolate chips. Serve immediately or frozen.

## Cranberry Nut Bread

2 cups all-purpose flour
1 cup sugar
1-1/2 teaspoons baking powder
1/2 teaspoon baking soda
1/2 teaspoon salt
1/4 cup butter or margarine
1 egg
1 tablespoon grated orange rind
3/4 cup orange juice
1-1/2 cups golden raisins
1-1/2 cups cranberries, chopped
1 cup walnuts or pecans, chopped

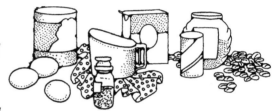

In a large bowl, sift together flour, sugar, baking powder, baking soda and salt. Then cut in butter or margarine until the mixture is crumbly. Stir in egg with orange rind and juice. Then blend this mixture into the flour mixture. Stir just until flour is moistened. Fold in the raisins, cranberries and nuts. Spoon the mixture into a greased 9-1/2 x 5-1/4 inch loaf pan. Bake at 350° for 60 to 75 minutes or until a toothpick inserted comes out clean. Cool in pan for 5 minutes. Then remove loaf to a wire rack to cool. Store well wrapped for 24 hours before slicing.

# Cranberry Upside-Down Muffins

*1 cup cranberries, cleaned and cut in half*
*1/2 cup butter or margarine, melted*
*1/2 cup firmly packed light brown sugar*
*2 cups all-purpose flour*
*1 tablespoon baking powder*
*2 tablespoons sugar*
*1/2 teaspoon salt*
*1/4 cup raisins*
*1 egg*
*1 cup milk*

Divide cranberries evenly into 12 greased or non-stick coated 3-inch muffin pans. Spoon about 1 teaspoon of the butter into each muffin tin. Sprinkle evenly with brown sugar. In mixing bowl, stir together flour, baking powder, granulated sugar, salt and raisins. In a smaller bowl, add egg, mix in the milk. Add to dry ingredients until moistened. Spoon batter over cranberries in prepared muffin pans, dividing evenly. Bake at 400° until golden brown, about 30 minutes. Let muffins stand in pan or on a wire rack for a few minutes or until bubbling stops. Then turn muffins out carefully and serve upside-down. Makes 12 muffins.

## Cranberry Walnut Bars

**Bar crust as follows:**

*2-1/2 cups flour*
*1 cup cold margarine, cut into 1/2-inch pieces*
*1/2 cup confectioners sugar*
*1/4 teaspoon salt*

Preheat oven to 350°. Mix all the ingredients in a large bowl until the mixture resembles coarse crumbs. Then press firmly into a greased 15x10x1 inch baking pan and bake for 20 minutes or until golden brown. Remove from the oven.

**Cookie bar recipe:**

*4 eggs*
*1-1/3 cups light or dark corn syrup*
*1 cup sugar*
*3 tablespoons margarine, melted*
*2 cups fresh or frozen cranberries, thawed, coarsely chopped*
*1 cup walnuts, chopped*

In a medium bowl, beat eggs, corn syrup, sugar or margarine until well blended. Stir in the cranberries and walnuts. Pour over hot crust and spread evenly. Bake for 25 to 30 minutes or until set. Cool completely on a wire rack before cutting into bars. Makes 48 bars.

# Cranberry Crumb Bars

**Crust:**

1-1/2 cups all-purpose flour
1-1/2 cups quick cooking oats
3/4 cup packed light-brown sugar
1/2 cup walnuts, finely chopped
1/2 teaspoon salt
3/4 cup unsalted butter, melted
1 tablespoon cold water

**Filling:**

2-1/4 cup fresh cranberries
3/4 cup red raspberry jam or
    preserves

Preheat oven to 375°. Line a 9-inch square baking pan with foil, letting the ends extend above the pan on 2 sides. Crust: Stir first five ingredients in a medium bowl until blended. Add butter and toss with 2 spoons until evenly moistened. Mixture will be in small clumps. Press about 2-1/2 cups evenly over bottom of prepared pan. Bake until golden brown, about 10 minutes. Meanwhile, add the water to the remaining oat mixture and toss until evenly moistened. Mix filling ingredients in a small bowl. Drop spoonfuls of filling over the crust and spread into an even layer with the back of a spoon. Crumble oat mixture over the top. Bake until the crust is golden and the filling bubbles, about 40 minutes. Cool completely in pan or on a wire rack. Cut in 16 squares. Remove from the foil.

## Streusel Cranberry Pie

*1/3 cup butter, softened*
*3/4 cup sugar*
*1/4 cup brown sugar*
*1 egg*
*1/2 teaspoon vanilla*
*1 cup flour*
*1/2 teaspoon baking soda*
*1/4 teaspoon salt*
*2 cups fresh or frozen cranberries*
*1/2 cup walnuts, chopped*
*1 9-inch unbaked pie shell*

**Streusel Topping:**
*2 tablespoons flour*
*1/2 cup packed brown sugar*
*1/4 teaspoon nutmeg*
*1/2 teaspoon cinnamon*
*2 tablespoons butter*
*1/2 cup walnuts, chopped*

In a medium bowl, cream together butter, sugar and brown sugar. Add and beat in egg and vanilla. Sift together flour, baking soda and salt. Stir this mixture into the egg mixture. This will be very thick. Fold in the cranberries and walnuts. Spread this mixture carefully into the pie shell. Bake at 350° for about 50 minutes or until the pie tests done with a wooden pick. While the pie is baking, blend together streusel ingredients until it is crumbly. Sprinkle topping over the pie as soon as it is removed from the oven; place an inverted bowl over the pie and let steam 20 minutes. Serve warm or cooled.

## Cranberry Pie

9-inch prepared 2 crust pie pastry
1-1/2 cup cranberries
1 cup raisins
1 cup water
1 cup sugar
3 tablespoons flour
2 tablespoons butter
2 teaspoons vanilla
1 teaspoon grated orange rind
1/4 teaspoon cinnamon
1 cup walnuts, chopped

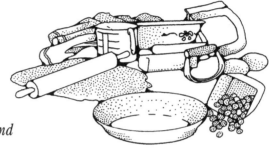

In a medium saucepan, boil together the cranberries and raisins in 1 cup of water until the skins pop. Mix the flour and sugar together and add to the cranberry mixture. Cook for a few minutes until slightly thickened. Remove from the heat and stir in butter, vanilla, orange rind, cinnamon and walnuts. Pour into pie shell. Adjust top crust. Bake at 400° for 15 minutes, turn oven temperature down to 350° and bake for 20 to 30 minutes more.

## Cranberry Pecan Pie

1 9-inch unbaked pastry pie shell
1 cup fresh cranberries, chopped
3 eggs
1 cup dark corn syrup
2/3 cup sugar
4 tablespoons butter or margarine, melted
1/2 teaspoon cinnamon
1/8 teaspoon nutmeg
1 cup pecan halves

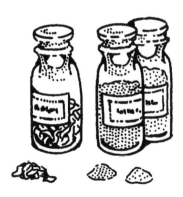

Preheat the oven to 325°. Prepare pastry shell and flute the edges as desired or press with fork. Sprinkle cranberries into the pie shell. In a bowl, beat eggs with the corn syrup, sugar, butter, cinnamon and nutmeg until mixture is well blended but not foamy. Pour over the cranberries. Carefully arrange the pecan halves in a series of circles over the filling. Bake for 50 to 55 minutes or until a knife inserted in the center comes out clean.

# RASPBERRIES

*No person will ever be a great leader who does not take real
joy in the success of the people under him.* Joan Bestwick

# Freezer Raspberry Sauce

*3 cups mashed fresh raspberries*
*3 cups sugar*
*1 cup light corn syrup*
*1-3 ounce package liquid fruit pectin*
*2 tablespoons lemon juice*
*4 cups whole fresh raspberries*

Combine the mashed raspberries, sugar and corn syrup, stirring until well mixed. Let stand for 10 minutes. In a small bowl, combine the pectin and lemon juice. Stir into the fruit mixture and mix for 3 minutes to distribute pectin evenly. Add the remaining whole berries, stirring carefully to distribute the fruit but leaving the berries whole. Ladle into 1-pint freezer containers, seal and let stand at room temperature for 24 hours or until partially set. Store in the refrigerator up to 3 weeks or in the freezer for 1 year. Thaw and stir before using. Serve over ice cream, sponge cake, shortcake waffles, etc.

# Raspberry Vinegar

*3 cups fresh raspberries*
*4 cups white wine vinegar*
*1/2 cup sugar*

Place berries in a 6 cup jar and set aside. In a medium saucepan combine vinegar and sugar. Bring almost to a boil over low heat, stirring constantly until sugar melts. DO NOT BOIL. Pour hot vinegar mixture over the berries, cover jar tightly and let stand at room temperature for 48 hours. Strain through several layers of cheesecloth into clean bottles or a jar. Seal tightly with a cork or lid. Store in a cool, dark place. Makes 4 cups.

# Raspberry Mandarin Soup

*2-11 ounce cans mandarin oranges*
*2 cups apple juice*
*1 cup orange juice*
*1/2 cup grape juice*
*1/2 cup lemon juice*
*1/4 cup quick cooking tapioca*
*2-10 ounce packages frozen raspberries in syrup*
*whipped topping*

Drain oranges, reserving the syrup. In a 2-quart saucepan, combine mandarin syrup, juices and tapioca. Let stand for 5 minutes then bring to a boil over medium heat for 5 minutes, stirring occasionally. Remove from heat; stir in raspberries until thawed. Add oranges and chill for 8 to 12 hours. Serve garnished with whipped topping.

## Raspberry Spinach Salad

*2 tablespoons raspberry vinegar*
*2 tablespoons raspberry jam*
*1/3 cup vegetable oil*
*3/4 pound fresh spinach, rinsed well and drained*
*3/4 cup whole pecans, toasted, divided*
*1 cup raspberries, divided*
*2 large kiwi, peeled and sliced 1/4 inch thick*

To make dressing, with a whisk or blender, blend vinegar and jam. Add oil in a thin stream while whisking. Set aside. To make the salad, mix spinach, half of the nuts, half of the raspberries and dressing. Mix before serving, then top with remaining nuts, berries and kiwi. 4 to 6 servings.

# Lettuce with Raspberry Dressing

*6 cups leaf lettuce, washed, drained and chilled*
*1/2 cup walnuts, toasted and coarsely chopped*

**Dressing:**
*1/3 cup vegetable oil*
*3 tablespoons sugar*
*2 tablespoons raspberry vinegar*
*1 tablespoon dijon mustard*
*1/2 cup fresh raspberries*

Whisk together all the dressing ingredients except for the raspberries. After the dressing is well mixed, fold in the raspberries. Refrigerate covered for at least 1 hour. Place lettuce in a glass bowl and add the walnuts. Toss with dressing. Serve immediately. Makes 6 servings.

# Raspberry Rice Salad

*1 cup rice, uncooked*
*3 cups milk*
*1 cup sugar*
*2 cups whipping cream, whipped*
*20 ounces frozen raspberries*
*2 tablespoons cornstarch*

Add milk and sugar to rice. Cook slowly for about 25 minutes, until milk is absorbed. Cool. Stir in whipped cream. Drain raspberries reserving the juice. Add cornstarch to juice and beat until smooth. Add raspberries and cook 3 more minutes. Cool and spoon raspberry sauce over the rice.

## Luscious Gelatin Fruit Salad

*2-3 ounce packages raspberry gelatin*
*1 envelope unflavored gelatin*
*1 cup boiling water*
*2 cups cold water*
*1-20 ounce can crushed pineapple with juice*
*2 large ripe bananas, mashed*
*1 pint fresh or frozen unsweetened raspberries*
*1 cup sour cream*

In a large mixing bowl, combine gelatins. Add boiling water and stir until dissolved. Add the cold water, then pineapple, banana and raspberries. Stir. Pour half of the gelatin mixture into a glass serving bowl, or 13x9x2 inch baking pan. Chill until firm. Let the remaining half of the gelatin sit at room temperature. Spread the sour cream evenly over the top of the firm, refrigerated gelatin mixture, then carefully pour the remaining mixture over the sour cream. Chill until firm.

# Frozen Raspberry Squares

*1-1/2 cups granola*
*1 quart vanilla frozen yogurt, softened*
*1 cup frozen whipped topping*
*1 cup raspberries, slightly mashed*
*1/2 cup chocolate syrup*

Lightly spray a 9-inch baking pan with non-stick cooking spray. Sprinkle the granola evenly over the bottom of the pan, breaking up any clumps. Spoon yogurt over the granola. Place in freezer. In a small bowl, place whipped topping and mashed raspberries, stir. Remove pan from freezer. Spread topping mixture over yogurt, cover and freeze for about 3 hours. To serve, drizzle each serving with chocolate syrup.

## Chocolate Dessert with Raspberry Sauce

*16-1 ounce semisweet chocolate squares*
*2/3 cup butter*
*5 eggs*
*2 tablespoons sugar*
*2 tablespoons all-purpose flour*

**Sauce:**
*2 cups fresh or frozen, whole unsweetened raspberries*
*1-3/4 cup water*
*1/4 cup sugar*
*4 teaspoons cornstarch*
*whipped cream*
*fresh whole raspberries to garnish*

Line the bottom of a 9-inch spring form pan with parchment paper; set aside. Place chocolate and butter in the top of a double boiler. Bring water to a boil; reduce heat and stir chocolate and butter until smooth. Add chocolate mixture to eggs, beating at medium speed about 10 minutes. Blend in sugar and flour, just until mixed. Pour into prepared pan. Bake at 400° for 15 minutes (cake will not be set in the middle) chill. For the sauce, combine raspberries, water and sugar in a saucepan. Bring to a boil. Reduce heat and simmer uncovered for 30 minutes. Put the mixture through a sieve and discard the seeds. Add water if needed to make 2 cups of juice. Combine cornstarch and 1 tablespoon water in a small bowl. Stir until smooth. Add cornstarch mixture to raspberry mixture. Cook over medium heat, stirring constantly, until mixture comes to a boil, cook and stir 1 minute longer. Remove from heat, and cool. To serve, spoon about 2 tablespoons sauce on each dessert plate, place a thin wedge of chocolate dessert on top of the sauce. Garnish with whipped cream and raspberries. Serves 16 to 20, makes 2 cups of sauce.

# Raspberry Custard Meringue

### Meringue:

*4 egg whites*
*1/2 teaspoon salt*
*1/4 teaspoon cream of tartar*
*1 cup sugar*
*1 teaspoon vanilla extract*
*1 cup finely chopped walnuts*

### Custard:

*1/4 cup butter or margarine*
*1/4 cup all-purpose flour*
*3-4 egg yokes*
*2 cups milk*
*1-1/2 teaspoons vanilla extract*
*1/8 teaspoon salt*
*3/4 cup sugar*

## Raspberry Custard Meringue, Cont.

**Topping:**
*8 ounces whipped cream*
*1/3 cup semisweet chocolate chips*
*3 tablespoons butter or margarine*
*1 quart fresh raspberries*

For the meringue, beat the egg whites, salt, and cream of tartar until soft peaks form. Add sugar, 1 tablespoon at a time, beating until whites are thick and glossy. Add the vanilla and fold in the nuts by hand. To form the crust, place meringue mixture on parchment paper; spread in a 14-inch diameter circle, or in several smaller circles for individual servings. Bake at 275° for 1 hour, turn off the heat and leave meringue in oven for at least 1 hour longer. Cool slowly. For custard, melt butter and add flour to make a paste. Wrap and cool completely. This mixture can be made the day before. Just before serving, add whipped cream and sweeten to taste. Melt the chocolate chips with butter in a small saucepan. Spoon custard mixture on meringue. Top with whipped cream and fresh berries. Drizzle with chocolate mixture.

## Black Raspberry Dumplings

*1 quart fresh or frozen black raspberries*
*1-1/4 cup sugar, divided*
*1 cup water*
*3 tablespoons cornstarch*
*3 cups prepared baking mix*
*1 cup milk*
*dash of nutmeg*

In a 6-quart pan, combine raspberries, 1 cup sugar, water and cornstarch, stirring to blend. Bring to a boil, stirring often. Turn heat to low. Meanwhile, combine baking mix, milk and remaining sugar, mixing until soft dough forms. Drop by spoonfuls onto boiling berries. Cook over low heat, uncovered, 10 minutes more, or until dumplings are cooked through. For glazed effect, sprinkle dumplings with additional sugar and a dash of nutmeg before serving. Makes 10 servings.

# Raspberry Streusel Muffins

## Muffins:
*1-1/2 cup all-purpose flour*
*1/2 cup sugar*
*2 teaspoons baking powder*
*1/2 cup milk*
*1/2 cup butter melted*
*1 egg beaten*
*1 cup fresh or frozen whole unsweetened raspberries divided*

## Streusel Topping:
*1/4 cup chopped pecans*
*1/4 cup packed brown sugar*
*1/4 cup all-purpose flour*
*2 tablespoons butter melted*

In a large bowl, combine flour, sugar and baking powder. In a small bowl, blend milk, butter and egg. Stir milk mixture into flour mixture just until moistened. Spoon about 1 tablespoon batter into each of 12 greased muffin cups. Divide half of the raspberries among cups, place remaining batter into the muffin tins and top with remaining raspberries. For topping, combine ingredients until mixture resembles moist bread crumbs, sprinkle over muffins. Bake at 375° for 20 to 25 minutes or until gold brown. Let stand 5 minutes; carefully remove from pans. Makes 12 muffins.

## Raspberry Cobbler

*4 cups raspberries*
*2/3 cup sugar*
*1/2 teaspoon lemon juice*
*2 tablespoons butter*
*1-1/2 cup biscuit mix*
*3 tablespoons melted butter*
*1 egg, slightly beaten*
*1/2 cup milk*
*whipped cream or ice cream*

In a medium bowl, toss the raspberries sugar and lemon juice. In a buttered 10x6 baking dish place the berries. Dot with 2 tablespoons butter. Combine biscuit mix, melted butter, egg and milk. With a spoon, drop the dough over the fruit. Bake for 30 to 35 minutes. Serve warm with whipped cream or ice cream. Serves 6.

## Raspberry Upside-Down Cake

*1 package white cake mix*
*3 cups fresh raspberries*
*1 cup sugar*
*1-1/3 cup whipping cream*
*fresh raspberries and whipped cream*

Grease and flour two 9x9x1-1/2 inch round cake pans, set aside. Prepare cake mix according to the directions, except reduce the water to one cup. Pour batter into prepared cake pans. Sprinkle 3 cups berries over the batter, sprinkle with sugar and pour the cream on top. Bake in a 350° oven for 1 hour. Immediately invert on to serving platters. If you like, serve with whipped cream and more berries.

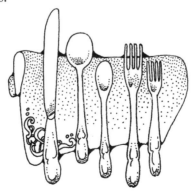

# Raspberry Crumble Coffee Cake

**Filling:**

2/3 cup sugar
1/4 cup cornstarch
3/4 cup water or raspberry juice
2 cups fresh or frozen raspberries
1 tablespoon lemon juice

**Cake:**

3 cups all-purpose flour
1 cup sugar
1 tablespoon baking powder
1 teaspoon salt
1 teaspoon cinnamon
1/4 teaspoon mace
1 cup butter, or margarine, softened
2 eggs, slightly beaten
1 cup milk
1 teaspoon vanilla extract

## Raspberry Crumble Coffee Cake, Cont.

**Topping:**
*1/4 cup butter or margarine*
*1/2 cup all-purpose flour*
*1/2 cup sugar*
*1/4 cup sliced almonds*

For filling: combine sugar, cornstarch, water or juice and berries. Cook over medium heat until thickened and clear. Add lemon juice. Set aside to cool. For cake: in a bowl, combine flour, sugar, baking powder, salt, cinnamon and mace. Cut in butter with a fork to form fine crumbs. Add eggs, milk and vanilla, stir until blended. Divide in half and spread into two round baking pans. Divide filling and spread evenly over batter by small spoonfuls over the batter and spread. For topping: cut butter into flour and sugar, stir in the nuts. Spread topping on cakes and bake at 350° for 40 to 45 minutes.

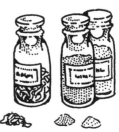

## Raspberry Coconut Pie

*1-10 ounce package frozen raspberries*
*1 envelope gelatin*
*1-8 ounce package cream cheese, softened*
*3 tablespoons confectioners sugar*
*1/2 pint heavy cream*
*1 prepared graham cracker crust*
*jelly beans, optional*
*1-4 ounce can flaked coconut*

Thaw raspberries slightly. Drain 1/2 cup juice into measuring cup. Sprinkle gelatin over the juice and stir. Combine softened cream cheese, confectioners sugar and raspberries in a bowl and beat with an electric mixer for 3 minutes, scraping the bowl frequently with spatula. Set cup of gelatin mixture in 1-inch of boiling water in a small saucepan. Stir until gelatin is dissolved, then gradually beat into the cream cheese mixture, then add heavy cream, beating until thick and fluffy, about 5 minutes, scraping bowl frequently. Turn into graham cracker crust. Decorate the pie with coconut and jelly beans.

# Cream Cheese Raspberry Pie

*6 ounces cream cheese softened at room temperature*
*1 teaspoon vanilla extract*
*3/4 cup heavy cream or whipping cream at room temperature*
*1/2 cup sugar*
*1-9-inch prepared graham cracker pie crust*
*1/2 quart raspberries*

In a small bowl, beat the cream cheese and vanilla together until well blended and light. Set aside. In a separate metal bowl, whip the cream until it doubles in volume. Continue beating on high speed, while adding the sugar a little at a time. Add the cream cheese mixture and beat again just briefly, just until combined. Pour mixture into the crust and cover with berries. Chill uncovered for at least 4 hours and serve.

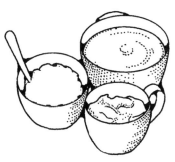

## Raspberry Apple Pie

*1 tablespoon lemon juice*
*2 cups apple, cored and thinly sliced*
*2 cups raspberries*
*1 cup sugar*
*2 tablespoons quick cooking tapioca*
*1/2 teaspoon salt*
*2 pastry pie crusts*
*2 tablespoons butter*

In a large bowl add fruit and lemon juice. Combine sugar, tapioca and salt. Add to fruit, toss lightly to combine. Let stand 15 minutes. Preheat oven to 425°. Line a 9-inch pie plate with bottom half of pie crust. Turn fruit into pie plate. Dot with butter. Top with remaining pie crust and seal the edges. Make several slits in the top of the pie crust. Bake for 40 to 50 minutes or until fruit is tender and crust is golden brown. Cool before serving.

# STRAWBERRIES

*What we practice proves what we profess.*

## Strawberry Syrup

*1 tablespoon cornstarch*
*1/3 cup water*
*1/4 cup sugar*
*2 cups sliced strawberries, 1 teaspoon juice*
*1 teaspoon lemon juice*

In a small saucepan, mix cornstarch and water. Add sugar and bring to a boil, stirring constantly until mixture is clear and thick. Add strawberries and lemon juice, heat through. Great on pancakes, French toast or waffles.

## Strawberry Float

*1/2 pint fresh strawberries, hulled*
*1/4 cup sugar*
*3 cups chilled milk*
*1/2 pint strawberry ice cream*

In a bowl, crush the strawberries with a potato masher and add sugar. With an electric mixer gradually add the milk. Spoon in ice cream and stir until it starts to melt. Serve in tall glasses.

## Strawberry Yogurt Dip

*1-1/2 cups fresh or frozen whole strawberries*
*1-8 ounce carton strawberry yogurt*
*1 cup whipped topping*
*Fruit or angel food cake*

Mash the berries in a bowl, add yogurt and mix well. Fold in whipped topping. Serve with fruit or cake.

## Lime Dip for Strawberries

*1-8 ounce carton sour cream*
*2 tablespoons powdered sugar*
*2 teaspoons finely shredded lime peel*
*1 tablespoon lime juice*
*3 cups strawberries*

In a small bowl stir the first four ingredients. Mix well. Place in serving bowl, chill. Serve with strawberries.

## Strawberries In Balsamic Vinegar
**(Great with poultry or seafood)**

*2 cups fresh strawberries, quartered*
*2 tablespoons sugar*
*1 teaspoon chopped mint*
*2 tablespoons balsamic vinegar*

Place the berries in a large bowl and sprinkle with the sugar. Add the mint and vinegar. Mix and let stand for 1 hour before serving.

## Strawberry Soup

*1 quart strawberries cleaned and halved*
*1/2 cup water*
*1/2 cup orange juice*
*1/2 cup sugar*
*2 cups yogurt*

Place all ingredients in a blender and mix well. Chill and serve. Serves six.

# Cinnamon Fruity Compote

*1-1/2 cup white grape juice*
*1-4-inch cinnamon stick*
*2 tablespoons sugar*
*1 teaspoon grated orange peel*
*1/4 teaspoon ground nutmeg*
*2 cups fresh strawberries, halved*
*3 medium nectarines, peeled and sliced*
*2 cups seedless grapes*
*1-15 ounce can mandarin oranges, drained*
*3 kiwi fruit, peeled and sliced*

In a small saucepan, bring the first five ingredients to a boil for five minutes. Remove pan from heat and let stand for 15 minutes. Remove the cinnamon stick and chill. Meanwhile in a glass bowl, layer 1/2 of the strawberries, all of the nectarines, grapes and oranges. Top with kiwi and remaining strawberries. Stir juice mixture and pour over fruit. Do not mix. Chill for 30 minutes. Makes 8 to 10 servings.

## Strawberry Rice Pudding

*3 cups water*
*1 cup white raw rice*
*1/2 teaspoon salt*
*1/3 cup plus 3 tablespoons sugar*
*1/2 teaspoon ground cinnamon*
*2 cups milk*
*1 teaspoon vanilla extract*
*1-1/2 pounds strawberries, sliced*
*1-8 ounce container vanilla yogurt*

In a medium-heavy saucepan, bring water to a boil. Add the rice and salt. Reduce the heat to medium-low and simmer uncovered until the rice is tender and the water is absorbed, stirring occasionally, about 20 minutes. Add 1/3 cup sugar, cinnamon, milk and vanilla, simmering until mixture is very thick, stirring frequently, about 25 minutes. Remove from heat and cool to room temperature. Mix strawberries and remaining 3 tablespoons of sugar in a bowl. Let stand until juices form, at least 30 minutes. Stir in yogurt and 2 cups strawberry mixture into the rice pudding. Transfer to large bowl. Top with remaining strawberry mixture and serve.

# Strawberry Spinach Salad

*6 cups (8 ounces) fresh spinach, torn*
*1/2 teaspoon toasted sesame seeds*
*2 cups fresh strawberries*
*1/4 cup salad oil*
*2 tablespoons red wine vinegar*
*1-1/2 teaspoons sugar*
*1-1/2 teaspoons fresh dill weed*
*1/8 teaspoon onion powder*
*1/8 teaspoon dry mustard*

In a salad bowl place spinach, sesame seed and strawberries, sliced in half. In a large jar with a top, combine remaining ingredients. Chill. Shake well before using on salad.

## Strawberry Peanut Salad

*3 medium apples, cubed*
*1-2 cups strawberries, sliced*
*1 cup celery, sliced thin*
*1/2 cup mayonnaise*
*3 tablespoons honey*
*3/4 teaspoon celery seed*
*3/4 cup salted peanuts*

In a large mixing bowl, combine apples, strawberries and celery. In a smaller bowl, combine mayonnaise, honey and celery seed, mix well. Just before serving, add peanuts to fruit mixture and drizzle with the dressing. Serves 6 to 8.

# Strawberry Salad

*1 large package strawberry gelatin*
*2 cups hot water*
*1-6 ounce package frozen strawberries, sliced*
*1 small can crushed pineapple with juice*
*2 bananas, sliced*
*1-8 ounce container sour cream*

Combine gelatin and boiling water until dissolved. Add all ingredients except the sour cream. Divide in half and congeal half in the refrigerator and then cover with sour cream. Pour the other half of the strawberry mix over the sour cream and congeal.

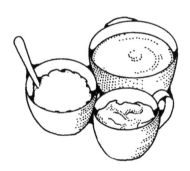

## Strawberry Pretzel Salad

*2 cups pretzels, crushed*
*3 tablespoons, plus 1/2 cup sugar*
*3/4 cup margarine, melted*
*1-8 ounce package cream cheese, softened*
*1-8 ounce container non-dairy whipped topping*
*1-6 ounce package strawberry gelatin*
*2 cups boiling water*
*20 ounces frozen strawberries*

Mix pretzels, 3 tablespoons sugar and melted butter into a 9 x 13 inch pan. Bake at 350° for 10 minutes and let cool. Cream softened cream cheese, beat in 1/2 cup of sugar and fold in the non-dairy whipped topping. Pour over cooled crust and chill. Dissolve gelatin in boiling water, add thawed strawberries and pour on top of cream cheese layer and chill. Cut in squares when firm.

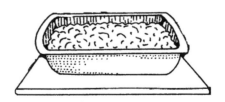

## Sour Cream Strawberries

*2 pint boxes strawberries, hulled*
*2/3 cup plus 6 tablespoons sour cream*
*1/3 cup plus 2 tablespoons light brown sugar packed*

In a large bowl slice strawberries. To this add 2/3 cup sour cream and 1/3 cup sugar, mix gently. Divide this mixture into 6 serving bowls. Top each with 1 tablespoon sour cream and 1 teaspoon brown sugar. Chill for at least 1 hour before serving.

## Strawberry Pancake Roll-Ups

*1-1/4 cup flour*
*1-1/2 cup milk*
*5 eggs*
*1/4 cup butter, melted*
*1/2 teaspoon salt*
*3 tablespoons sugar*
*3 cups strawberries, sliced*
*1/2 cup sugar*
*powdered sugar*

Make a batter from the flour, milk, eggs, butter, salt and 3 tablespoons sugar in a medium bowl. In a separate bowl, mix together strawberries and sugar. In a large skillet, melt 1 tablespoon butter and cook batter until golden brown before turning to brown other side. Keep pancakes warm. Place 1/2 cup strawberries on edge of each pancake and roll up. Sprinkle with powdered sugar. Serve with whipped cream or sour cream. Makes 6.

## Strawberry Blossoms

*8-12 large strawberries*
*3 ounce package cream cheese, softened*
*2 tablespoons powdered sugar*
*1 tablespoon sour cream*

Remove stems from berries forming a flat base. Place berries pointed end up. With a sharp knife, carefully slice berry in half vertically through center to within 1/4 inch of the base. Cut each half into 3 wedges forming 6 petals, do not cut through base. Pull petals apart slightly. In a small bowl, beat cream cheese, sugar and sour cream until light and fluffy. Fill berries with mixture.

## Strawberry Tarts

*6 baked tart shells, 3-1/2 inches in diameter*
*1-1/2 cups strawberries, hulled and sliced*
*1/4 cup sugar*
*1-1/2 cups whole strawberries, hulled*
*1/2 cup currant jelly*
*2 drops red food coloring*
*whipped topping*

In a bowl, combine sliced berries and sugar. Cut the whole berries lengthwise. In a small saucepan over low heat, stir jelly with food coloring until melted, cool. Just before serving tarts, assemble. Spoon about 3 tablespoons sliced berries into each tart shell. Top this with 4 or 5 berry halves, arrange in a pattern. Brush the strawberries with the jelly and top with whipped topping.

## Strawberry Squares

*1 cup all-purpose flour, sifted*
*3/4 cup brown sugar*
*1/2 cup chopped walnuts*
*1/2 cup butter, melted*
*2 egg whites*
*1 cup granulated sugar*
*2 cups strawberries, sliced*
*2 tablespoons lemon juice*
*1 cup whipped cream*

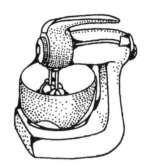

In a medium bowl, mix flour, sugar, walnuts and butter. Bake in a shallow pan at 350° for 20 minutes. Stirring occasionally. Sprinkle 2/3 of the crumbs in a 13x9x2-inch pan. Combine egg whites, granulated sugar, berries and lemon juice. Beat at high speed about 10 minutes. Fold in whip cream and spoon over crumbs. Top with remaining crumbs. Freeze for 6 hours. Serves 12.

## Strawberry Muffins

*1-1/2 cups strawberries, mashed*
*3/4 cup sugar, divided*
*1-3/4 cups flour*
*1/4 teaspoon nutmeg*
*1/4 teaspoon salt*
*1/2 teaspoon baking soda*
*2 eggs, beaten*
*1/4 cup butter*
*1 teaspoon vanilla*

Combine strawberries and 1/4 cup sugar; set aside. After 30 minutes drain strawberries, reserving the liquid. Combine flour, nutmeg, salt and soda, set aside. In a medium bowl, mix eggs, butter, vanilla, remaining 1/2 cup sugar and reserved liquid from the berries. Add to flour mixture, stir until combined. Fold in berries. Spoon into greased muffin tins. Bake at 425° for about 20 minutes. Makes 12 muffins.

## Strawberry Banana Nut Bread

*1 stick margarine, softened*
*3 eggs*
*1 cup sugar*
*1/2 pint fresh strawberries, chopped*
*3 medium ripe bananas, sliced*
*1/2 cup milk*
*1/2 cup nuts, chopped*
*3 cups flour*
*2 teaspoons baking powder*
*1/2 teaspoon salt*

In a mixing bowl, beat together margarine, eggs and sugar. Stir in strawberries and bananas. Add milk, nuts, flour, baking powder and salt. Divide between three greased loaf pans. Bake at 350° for 35 to 40 minutes. Cool.

## Strawberry Cheesecake Truffle

*2-8 ounce packages cream cheese*
*2 cups confectioners sugar*
*1 cup sour cream*
*1/2 teaspoon vanilla extract*
*1/4 teaspoon almond extract*
*1/2 pint whipping cream*
*1 teaspoon vanilla extract*
*1 tablespoon sugar*
*1 angel food cake, torn into bite size pieces*
*2 quarts fresh strawberries, thinly sliced*
*3 tablespoons sugar*
*3 tablespoons almond extract*
*large glass bowl for serving*

In a large bowl, cream together cream cheese and sugar. Add sour cream, vanilla and 1/4 teaspoon almond extract. Set aside. In a small, deep bowl, whip the cream, 1 teaspoon vanilla and sugar. Fold whipped cream into cream cheese mixture. Add cake pieces and set aside. Combine strawberries, sugar and almond extract. In a large glass bowl, layer together: strawberries, cake mixture, strawberries, cake mixture and finish with strawberries. Cover and chill well. Makes 24 servings.

## Strawberry Nut Cake

**Cake:**
*1 box white cake mix*
*1 box strawberry gelatin*
*4 eggs*
*1 cup milk*
*1 cup oil*
*1 cup frozen strawberries*
*1 cup flaked coconut*
*1 cup chopped nuts*

**Icing:**
*1 box powdered sugar*
*1 stick butter*
*1/2 cup strawberries*
*1/2 cup chopped nuts*
*1/2 cup coconut*

Cake: pour all dry ingredients into a mixing bowl. Add milk, eggs and oil. Mix until smooth and stir in the strawberries, coconut and nuts. Bake at 350° in three 8-inch baking pans for 25-30 minutes.
Icing: mix powdered sugar and butter together until blended; then add the strawberries, nuts and coconut. Stir until well blended. Frost cake.

# Strawberry Pound Cake

**Cake:**
*1 box white cake mix*
*1 cup fresh strawberries, crushed*
*1-3 ounce box strawberry flavored gelatin*
*1/2 cup vegetable oil*
*4 eggs at room temperature*

**Glaze:**
*1/4 cup butter or margarine*
*3-1/2 cups confectioners sugar*
*1/4 cup strawberries, mashed*

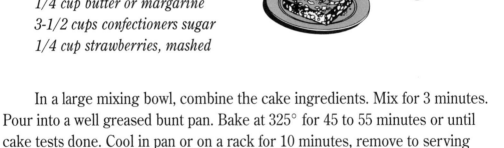

In a large mixing bowl, combine the cake ingredients. Mix for 3 minutes. Pour into a well greased bunt pan. Bake at 325° for 45 to 55 minutes or until cake tests done. Cool in pan or on a rack for 10 minutes, remove to serving plate. Combine glaze ingredients until smooth and spread over the cake.

## Strawberry Cake

*1 cup flour*
*1/2 cup sugar*
*2 teaspoons baking powder*
*dash of salt*
*1 egg*
*2 tablespoons butter, melted*
*1/2 cup milk*
*1-1/2 cups strawberries, sliced*

Sift together dry ingredients. Add the egg, butter and milk and beat for two minutes. Pour into a buttered 8-inch square pan. Top with strawberries. Sprinkle with a topping made of:

*1/2 cup flour*
*1/2 cup sugar*
*1/4 cup butter*
*1/4 cup walnuts, chopped*

Mix these ingredients into a crumb mixture. Bake at 375° for 35 minutes.

## Unbelievable Strawberry Pie

**Pie crust:**

*1-1/2 cups flour*
*1/2 teaspoon salt*
*2 tablespoons sugar*
*1/2 cup vegetable oil*
*2 tablespoons cold water*

**Filling:**

*11 ounces cream cheese, room temperature*
*4 tablespoons sugar*
*1 quart fresh strawberries, divided*
*2 ounces semisweet chocolate*
*1 tablespoon nuts, finely chopped*

Combine all of the crust ingredients in a 9-inch pie plate, mix and press into the bottom and sides of the plate. Bake at 400° for 12 to 15 minutes or until golden brown. Cool. Beat together cream cheese and sugar with mixer until smooth. Add about 3/4 cup of the strawberries; beat until just bits of berry remain. Spread mixture into cooled pie shell. Melt chocolate over low heat, dip tips of remaining strawberries into chocolate. Arrange, tips up, over cream cheese layer. Sprinkle with nuts for garnish. Chill thoroughly and cut. Serves 8.

# MIXED BERRIES

*From small beginnings come great things.* Proverbs

## Cherry-Strawberry Jam

*1-1 pound, 4 ounce can pitted, tart, red cherries, water packed*
*1-10 ounce package frozen strawberries, thawed and sliced*
*4-1/2 cups sugar*
*3 tablespoons lemon juice*
*3 ounces liquid fresh fruit pectin*

Drain cherries, reserving the juice, and chop. Measure and add enough reserved juice to make 2 cups. In a large saucepan, combine the fruits, sugar and cherry and lemon juice. Bring to a boil and boil hard for 1 minute stirring constantly. Remove from heat and stir in pectin at once. Skim off the foam. Stir and skim for 5 minutes. Ladle quickly into hot scalded jars and seal. Makes 6-1/2 pint jars.

## Rasp-Blue Jam

*1 quart red raspberries*
*1 pint blueberries*
*7 cups sugar*
*1 bottle liquid fruit pectin*

In a bowl, crush berries to measure 4 cups. In a kettle add berries and sugar. Mix well. Heat to a rolling boil, boil hard for one minute, stirring constantly. Remove from heat, stir in pectin and skim off the foam. Ladle into hot sterilized jar. Seal. Makes 5 pints.

## Cranberry-Pineapple Sauce

*1-12 ounce bag fresh or frozen cranberries, thawed*
*1-8 ounce can crushed pineapple, or tid bits in juice*
*1 cup sugar*
*1/4 cup orange juice*
*1 teaspoon grated orange rind*
*2 teaspoons vanilla*

In a food processor or blender, coarsely chop cranberries and the pineapple with its juice, 3/4 cup sugar and orange juice. When this is done, pour this mixture into a medium saucepan. Bring this to a boil over medium heat. Reduce the heat and simmer, stirring occasionally until thickened, about 10 to 12 minutes. Add remaining 1/4 cup of sugar and remove this from the heat. Stir in the rind and vanilla. Transfer to container and cool. Refrigerate to chill. Makes 12 servings.

## Blackberry-Raspberry Sauce

*1/2 cup sugar*
*1/3 cup water*
*1 cup blackberries*
*1 cup raspberries*
*1 tablespoon fresh lemon juice*

In a small saucepan, bring sugar and water to a boil. Stirring until sugar is dissolved. Place in a blender and add berries and lemon juice. Pulse until berries are chopped. Cool sauce before serving.

## Cran-Strawberry Sauce

*1-10 ounce package frozen strawberries in syrup, thawed and sliced*
*1-12 ounce package of fresh or frozen cranberries*
*1/2 cup sugar*

Drain strawberries, reserving the syrup. Measure the syrup and add enough water to equal 1 cup. In a medium saucepan, over high heat, combine syrup with cranberries and sugar. Bring to a boil then reduce heat to low. Cook stirring until cranberries burst and mixture thickens about 15 to 18 minutes. Remove from heat. Stir in the reserved strawberries. Refrigerate until completely chilled at least 3 hours or overnight. Makes three cups.

## Fresh Fruit Soup

*2 cups water*

*2 tablespoons quick cooking tapioca*

*1-6 ounce can frozen orange juice concentrate, thawed*

*2 tablespoons sugar*

*1 tablespoons honey*

*1/8 teaspoon almond extract*

*2-1/2 cups of fresh fruit like blueberries, raspberries, strawberries, grapes,*
    *kiwi, peaches*

In a 2-quart saucepan, combine water and tapioca. Let stand for 10 minutes. Bring to a boil. Boil for 2 minutes, stirring constantly. Remove from heat and stir in the orange juice concentrate, sugar, honey and extract. Add fresh fruit, chill and serve.

## Triple Berry Shake

*1 pint fresh strawberries, cleaned and halved*
*1/2 cup skim milk*
*1/4 cup fresh blueberries*
*1/4 cup fresh raspberries*
*1/2 cup small ripe banana, sliced*
*2 tablespoons honey*
*1 cup ice cubes*

In a blender, add all ingredients and blend until well mixed and foamy. Serve.

# Berry Orange Grunt

**Fruit:**

4 cups berries (blueberries,
   raspberries or blackberries)
1 cup sugar
1/4 cup orange juice concentrate
1 teaspoon grated lemon rind

**Dumplings:**

1 cup all-purpose flour
1/4 cup granulated sugar
2 teaspoons grated orange peel
1 teaspoon baking powder
1/2 teaspoon baking soda
1/4 teaspoon salt
1/4 teaspoon ground nutmeg
2 tablespoons unsalted butter, melted
3/4 cup buttermilk

Fruit: Mix berries, sugar, orange juice concentrate and rind in a 9-10 inch skillet with a tight fitting lid. Simmer, uncovered for 2 to3 minutes.

Dumplings: Mix flour, sugar, peel, baking powder, baking soda, salt and nutmeg in a large bowl. Mix melted butter and buttermilk in a small bowl. Add to flour mixture, stirring to moisten and form a loose dough. Drop dough over fruit in small dumplings. Cover tightly; cook over medium heat for 20 minutes, until dumplings are set. After 15 minutes, check for dumpling firmness. NO PEEKING BEFORE THEN. To serve, scoop dumplings into dishes. Cover with sauce and serve with ice cream if desired.

## Blueberry-Strawberry Buckle

*1/2 cup butter at room temperature*
*1 cup sugar*
*1 egg*
*1 teaspoon vanilla*
*1 cup and 1/4 cup flour*
*1 teaspoon baking powder*
*1 teaspoon salt*
*1/2 cup milk*
*1 cup blueberries*
*1 cup sliced strawberries*
*1/2 teaspoon cinnamon*
*1/2 teaspoon nutmeg*
*whipped topping or ice cream*

Preheat oven to 350°. Grease a 9-inch square baking pan. In a small bowl, cream 1/4 cup butter and 1/2 cup sugar. Blend in the egg and vanilla. In a small mixing bowl, stir together 1 cup of the flour, baking powder and salt. Add dry ingredients to creamed mixture alternating with milk. Stir until well blended. Pour batter into prepared pan. Arrange fruit over the batter. In a small bowl, combine remaining sugar, flour, cinnamon and nutmeg. Cut in remaining butter until mixture is crumbly and sprinkle mixture over fruit. Bake for 35 minutes and serve.

## Black and Blue Crumble

*3 cups fresh blueberries*
*3 cups fresh blackberries*
*1 tablespoon fresh lemon juice*
*8 tablespoons granulated sugar, divided*
*3/4 cup plus 4 tablespoons flour, divided*
*1/2 cup firmly packed brown sugar*
*1/2 teaspoon cinnamon*
*6 tablespoons butter or margarine, cut into small pieces*
*1/2 cup walnuts, chopped*
*vanilla ice cream*

Heat oven to 375°. Combine blueberries, blackberries, lemon juice, 6 tablespoons granulated sugar and 2 tablespoons flour in a large bowl. Transfer to a 2-quart shallow baking dish. Combine remaining flours and granulated sugar, brown sugar and cinnamon in a medium bowl with a pastry blender. Cut in butter until crumbly. Stir in walnuts. Sprinkle over fruit. Bake 40 to 45 minutes, until top is brown and fruit is bubbly. Cool on wire rack.

# Fresh Fruit Trifle

*1 package yellow or white cake mix*
*fresh fruit, strawberries, blueberries, etc.*
*1 cup vanilla pudding*

Prepare and bake cake according to package directions. When cool, cut into large bite size pieces. Add the fruit, including the juice and pudding to the cake, mix well. Chill for several hours. Can be served with non-dairy whipped topping.

# American Cheesecake

*1 9-inch pie crust*

**Berry layers:**

*1-1/2 cups sugar*
*4-1/2 tablespoons cornstarch*
*1-1/2 cups water*
*4-1/2 tablespoons raspberry-flavored gelatin powder*
*1 pint fresh or frozen blueberries*
*1 teaspoon fresh lemon juice*
*1 pint fresh or frozen raspberries*

**Cream layer:**

*4 ounces cream cheese, at room temperature*
*1/3 cup confectioners sugar*
*4 ounces nondairy frozen topping, thawed*

For berry layers, combine sugar, cornstarch and water in a medium saucepan, stirring to dissolve. Cook until thick and clear. Add gelatin and stir until dissolved. Divide mixture in half. Stir blueberries and lemon juice into half of the mixture. Spread over the bottom of the pie shell and refrigerate. Fold the raspberries gently into the remaining half of the mixture and set aside. For cream layer, beat together the cheese and sugar until smooth. Mix in the topping and spread over the blueberry layer. Refrigerate until set. Carefully spread the raspberry layer over the cream cheese layer. Chill for at least 4 hours before serving.

## **Summer Fruit Pie**

*1 8-inch prepared graham cracker pie crust*
*2 tablespoons apple jelly*
*1 tablespoon water*
*1/2 cup grapes, halved*
*1/2 cup blueberries*
*1/2 cup strawberries, halved*
*2 fresh peaches, sliced*

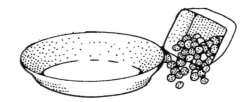

In a small saucepan, stir together jelly and water. Cook over low heat for 3 minutes. Arrange fruit over baked crust. Drizzle jelly mixture over fruit. Serve either chilled or warm.

## R&B Pie

*1 10-inch graham cracker pie shell*
*1 pint whipping cream*
*3 tablespoons powdered sugar*
*1 teaspoon almond extract*
*1 pint raspberries*
*1 cup blueberries*

In a deep bowl, whip cream until thick. Add powdered sugar and extract, blending well. Spread on the pie shell. Arrange the raspberries around the edge and mound blueberries into the center. Chill and serve.

# INDEX

*Let God love you through others and
let God love others through you.* D.M. Street

ISBN 0-932212-94-8

Look for Joan Bestwick's *Life's Little Zucchini Cookbook, Life's Little Rhubarb Cookbook, Life's Little Peaches, Pears, Plums & Prunes Cookbook, Life's Little Apple Cookbook* and *Life's Little Blueberry Cookbook* also by Avery Color Studios, Inc.

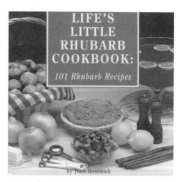

ISBN 1-892384-00-0

ISBN 1-892384-11-6

ISBN 1-892384-22-1

ISBN 1-892384-32-9

Avery Color Studios, Inc. has a full line of Great Lakes oriented books, puzzles, cookbooks, shipwreck and lighthouse maps and lighthouse posters.

For a full color catalog call: **1-800-722-9925**

Avery Color Studios, Inc. products are available at gift shops and bookstores throughout the Great Lakes region.